# Galois Fields in
# Quantum Mechanics

## Ed Gerck, Ph.D.

Formerly with the Max-Planck Institute for Quantum Optics
CEO and Founder of Planalto Research

Published by Planalto Research
Mountain View, CA, USA

Gerck, Ed.
Galois Fields in Quantum Mechanics

# PREFACE

Galois fields are applied in mathematics, cryptography, error-correction codes, and software, mainly to enable the usual four operations (+ - x ÷) of arithmetic, as when using the cumbersome real numbers, but using only integers, *exactly*, and *fast* in computers. Calculations become no longer approximate. Not designed for human use, the ideas behind a Galois field are easily codeable.

They are not yet the established art, or even applied today, in quantum mechanics or relativity, and that is due to a number of factors, including the lack of tools in calculus, which depends on continuity to exist (a derivative existence implies continuity). With Galois fields, there is no infinitesimal, Cauchy epsilon-delta formalism, Dedekind cuts, and other tools as used in analysis.

**That changes with this book.** We introduce all that one needs to know to use Galois fields in quantum mechanics, special and general relativity, and their novel union, including calculus without derivatives --- we use difference-order equations. <u>This is not an approximation.</u>

We are lifting the roadblock of continuity in calculus by pointing out that continuity does not even exist, and using the mathematical theory of Digital Constructivism (DC).

Ed Gerck
Mountain View, California

# To the skepticals

**Why change quantum mechanics?** Many do not believe that seemingly "esoteric" and abstract structures in pure mathematics, as rings and fields, important as they are in modern algebra, should be relevant in physics, applied mathematics, or even in mathematics. Today, though, to highlight their importance, there are already hardware and software libraries that make it "plug-and-play" to use Galois Fields, for the four operations (+ - x ÷) of arithmetic.

But we may need to calculate what current Galois fields theory say that we cannot, such as in calculus and renormalization (how to deal with infinities), and do not take into account things that cannot be measured, like the experimental impossibility of the measurement of infinitesimally small distances, the basis of the assumptions of continuity and uniformity in current physics, as in quantum field theories (QFT)[1], such the Standard Model and General Relativity.

---

[1] Today, the Standard Model of particle physics, which accounts for all observed phenomena on length scales **larger** than ~ $10^{-18}$ meters, is a QFT. There are no "particles" in the Universe, or classical Fourier waves, just quantum waves -- and QFT models this behavior. The "particles" are seen as vibrations of the fields, where the fundamental entities are not the "particles", but the field of each type (e.g., photon, electron, whatever) as an abstract object that penetrates spacetime in QFT, for the whole universe. General relativity can also be viewed as a QFT, or an 'effective quantum field theory' describing the long-distance behavior of a massless spin 2 "particle", the graviton, as it interacts with the "particles" from the rest of the SM.

**How to change quantum mechanics?** By showing first that "continuity" does not exist, so that it is not even considered, and students can move faster with less contradictions. The hypothesis of "continuity" is no longer a limitation in calculus or physics.

We also motivate that software development and quantum mechanics are as difficult as they still are because people have developed the "right" intuitions for mental processes to be consistent with classical (Newtonian) mechanics, using "continuity" because this "sounds" familiar — but it does not. Nothing is continuous in Nature nor mind, mass is not conserved in a closed system, and time is not absolute.

Yes, mathematics is only one. And all of mathematics is built from 1 and counting, even zero, even Galois fields. But, there are many philosophical schools in mathematics. In our treatment of Galois fields we follow Digital Constructivism, or DC, where one uses constructive mathematics (itself subdivided in various schools of thought) as distinguished from its traditional counterpart, classical mathematics, by the strict interpretation of the phrase "there exists" as "we can construct".

Mathematics, Physics, and software, including all of calculus, can then be rewritten on these new terms. **No continuity is required nor exists.**

# Reality is

# Discrete and Finite

Abstract

We formulate in quantum mechanics a variational principle in all coordinates of relativistic spacetime in equal footing as fractal dimensions, with entanglement, and consider an abstract, discrete field $f(x)$ with a topology given by the finite rational numbers, all that one can measure physically, which permeates spacetime. Next, we consider vibrations of this discrete field in all eigenmodes concurrently, within the entire boundary conditions, and introduce its piece-wise series as the desired linear combination of eigenfunctions $f_k(x)$. This reduces to the time-independent Schrödinger equation for bound states in 3 dimensions of space (without using derivatives or continuity), which is validated specifically for four major potential models: harmonic, Coulomb, linear, and of Rydberg states. This can be seen as the same propagation onward in a logistic map in chaos theory, connecting both spaces, and offering a quantum origin for Feigenbaum's constant.

## I.  INTRODUCTION

According to the Hubble law [1], the age of the universe is *ca.* 13.8 billion years [2, 3]. The universe is thus finite, and the number of stars is finite, made by finite actions in a finite duration [1].

Quantum mechanics (QM) is well-known to be in a total conceptual rupture with classical mechanics, and with the concepts of continuity, sub-summed here as the many aspects and needs artificially imposed by *'relativistic field-continuity, uniformly varying variables (uniformity), local causality, infinity, $(\epsilon - \delta)$ Cauchy formalism, determinism, and renormalization'* (hereafter called figuratively the

--------

[1] Objects may combine or split, and change type, but remain finite.

'7 Plagues' of physics), that we eschew, and therefore eliminate the requirements that give origin to the '7 Plagues' — which seem all to be traceable to the false idea of continuity.

Yet QM follows the classical field-continuity, which remained the basis of quantum field theories (QFT) [2], and this in turn require renormalization and others of the '7 Plagues' to deal with fields that have, e.g., modes of arbitrarily short wavelength or time — as if that would be real.

According to QM, Nature jumps – and mitosis/meiosis is an example in biology of this quantum behavior

---

[2] Where a quantum field is, unfortunately, considered a superposition over classical fields.

in life, as well as of the possible separation stages. In QM, the energy field must have a discontinuous, discrete structure [3].

We formulate here a consistent QM framework, albeit without any continuity hypothesis. There is nothing probabilistic about an encrypted text using Galois fields. The decryption is a certain, unique text, not a probabilistic range.

We can only measure finite integers, and their ratio as a finite rational; so-called real numbers are used for their nice field properties, but they are not actually real — the reality that we can see and

---

[3] The discontinuity is often described to mean that between two points there is a nothing — no objects, no atoms, no molecules, no particles, just nothing, where even the word 'nothing' is maybe too much.

measure is limited to finite rationals, which can be mapped to integers – the quantum found by Max Planck.

Leon Brillouin, ca. 1956, also concluded that the infinitely small does not exist. We confirmed the prevalence of Galois fields over so-called real numbers. This opens up to work on new results including software, the scalability of quantum computing, and the unification of general relativity with quantum mechanics.

We used a variational principle in all coordinates of relativistic spacetime, as fractal dimensions, which considers an abstract, discrete field $f(x)$ with a topology that permeates spacetime. This new framework reduces to the time-independent Schrödinger equation for bound states (without using derivatives, which eschews continuity), validated in 3D specifically for four major potential models: harmonic, Coulomb, linear, for Rydberg states, and in general.

This is seen next as the same propagation onward in logistic (iterated) maps in chaos theory [4, 5], connecting both spaces, and offering a quantum origin for the Feigenbaum constant. Not using continuity preempts, e.g., using renormalization and to explain universality in QM, and can be used to compute Feigenbaum's constants.

## II. DISCRETE QUANTUM FIELD THEORY (DQFT)

We begin with fundamental mathematics, to better define terms. We will use 'field' in different meanings, clear from context, but we must disambiguate terms first — in physics and mathematics, as well as in code, cryptography, and error-correction [6].

The meaning in physics is clear for the intended audience of this work. In mathematics, though, a 'field' is any set of elements that satisfies the field axioms for both addition and multiplication and is a commutative division algebra [4].

---

[4] An archaic name for a field is *rational domain*. The French term for a field is *corps* and the German word is *Körper*, both meaning *body* in English.

The group of integers modulo $p$, where $p$ is a prime number, is denoted in mathematics by $\mathbb{Z}/\mathbb{Z}p$. It is well-known that $\mathbb{Z}/\mathbb{Z}p$: (1) is an abelian group under addition; (2) is associative and has an identity element under multiplication; (3) is distributive with respect to addition, under multiplication; (4) is a mathematical field.

A mathematical field with a finite number of members is known as a finite field or *Galois field*. For each prime power, there exists exactly one (up to an isomorphism) finite field $GF(p^n)$, often written as $F_{p^n}$. The order $p$ of a finite field is always a prime or a power $n$ of a prime. The advent of 128-bit instructions, such as Intel's Streaming SIMD Extensions [7], allows us to perform Galois Field arithmetic much faster, in hardware. The paper [7] details the SIMD instructions to multiply regions of bytes by constants in $F_{2^w}$ for $w$ in $4, 8, 16, 32$.

Today, we also use $F_{((q^{n}1)^{n}2)^{n}3}$, and further, a more complex extension of the prime finite field F(q). The initial prime field F(q) used at the lowest level of the construct is frequently called the basic finite field with respect to the extension. This explanation should be thought throughout, to denote what will be written more simply as a set of the *finite natural numbers* — implicitly understanding integers, isomorphism, self-similarity, and fields in mathematics, and Galois fields of order $p$ and $n$ power, $F_{p^n}$.

By definition, any finite set for natural numbers end in number M of elements, which can be put (e.g., the odd numbers from 5 to 31 and skipping 7) in a 1:1 correspondence with the integers mod $p$, $p \leq M$, which is a field. Any finite set, not just of numbers, can become thus a field using this process. We denote this by $\mathbb{Z}/\mathbb{Z}p$, a field.

Finite fields as $F_{p^n}$ are used extensively in the study of cryptography and error-correcting codes [7]. We investigate next how QM could also be done in Galois fields, as a basic expression of quantum behavior.

Infinity will be used here, as well as the symbol $\infty$, meaning an unknown number in algebra, as high as wanted, which is not pre-determined, fixed (i.e., a number), but is finite, and a result that *cannot be counted.* In that way, infinity should be treated here as a *finite irrational number.*

Therefore, while no finite field is infinite in the usual sense, there are infinitely ($\infty$ as defined) and (e.g, isomorphic, self-similar) many different finite fields.

We next use an example of a first-order *difference* equation in discrete mathematics, over the field of finite natural numbers, understood here as the Galois field $F_{p^n}$. This uses only two points to represent *exactly* a second-order derivative, as presented first by the author [8], which operation is ubiquitous in physics.

The second-derivative is represented *exactly* in any function spanned by linear combinations in the set $U$:

$$U = \{e^{-\alpha x}, xe^{-\alpha x}, x^2 e^{-\alpha x}\}, \qquad (1)$$

with $\alpha > 0$ [8–11], which already obeys the known boundary conditions of the TISE (cf. Appendix A).

As explained further in Appendix A, the hypothesis of continuity between the two points was *not* assumed. Therefore, we were not pushing our equations to arbitrarily short distances. This would mean arbitrarily high energy, where they, probably, don't actually apply.

Thus, at *two* points,

$$\frac{d^2\psi(x)}{dx^2} = t\psi(x) - p\psi(x - q), \qquad (2)$$

where $t = \alpha^2$, $p = (2\alpha^2/e)$, $q = 1/\alpha$, and $\alpha = 1/r$, wherein $\alpha$ is a variational constant in the interval between the two points, and taken as $\alpha_k$. The distance thus between the two points $r_{k-1}$ and $r_k$ is $1/\alpha_k$, the variational choice for Eq.(2) to be given in two points [8].

The Time-Independent Schrödinger Equation (TISE) (see Appendix A) can then use Eq.(2) to be written without the hypothesis of continuity (as needed by the derivative). Normalizing in atomic units,

$$\psi(r_{k-1}) = \frac{e}{\alpha_k^2}[E - \alpha_k^2 - V_k]\psi(r_k), \qquad (3)$$

where the index $k = 1, 2, 3, ...K$ refers to the partition of the coordinate space, as usual, $\alpha_k$ is a piece-wise variational parameter, and the boundary conditions are $\psi(0) = \psi(\infty) = 0$, where $\infty$ (is defined as above) represents a large enough, finite, separation in spacetime [5].

The solutions are all in the form,

$$E_k = \alpha_k^2 + V(x_k) = E(an + b + c/n + \mathcal{O}(1/n^2),$$

$$(4)$$

where $a$, $b$ and $c$ belong to the set of finite rational numbers, and the correction order $\mathcal{O}$ tends to vanish as $1/n^2$ for large $n$ [8–10].

---

[5] Eq.(3) can be written in the bi-diagonal matrix form $M\psi = E\psi$ and trivially solved, where the eigenvalues are already located in the main diagonal.

This allows one to construct more than $K$ non-trivial linear combinations of piece-wise functions in the set $U$ that obey the boundary conditions, so that the general solution to Eq.(3) is, where each k-piece of $\psi(x)$ is a linear combination,

$$\psi(x) = (A_k + B_k x + C_k x^2)(e^{-\alpha_k x}), \qquad (5)$$

for $r_{k-1} \leq x \leq r_k$, for all $x$ and $k = 1, 2, 3, ...K$, and we assume that all possible solutions can co-exist.

Since $\psi(x)$ is a function of the variational parameters $\alpha_k$, the variational principle [10] for the quantum levels $E_n$ leads to,

$$E_n = \frac{1}{2}\alpha_n^2 + V(\frac{an + b + c/n + \mathcal{O}(1/n^2)}{\alpha_n}), \qquad (6)$$

16

where $n$ is the quantum number $n = 1, 2, 3, ...N$

For example, and where all the eigenvectors obey Eq.(3), so they include all the possible contributions of each of the $\alpha_k$, each $E_n$ eigenvalue below only includes their own, single $\alpha_n$ (here denoted as $\alpha$, for simplicity), we tested the following list of significant potential functions:

1. HARMONIC: For a harmonic potential $V(x) = Kx^2$, Eq.(4) gives:

$$E_n = \alpha^2, \tag{7}$$

which is *exactly* the usual functional dependence of the eigenvalues with the quantum number, reproducing all the known eigenvalues with $a = 2$ and $b = 1/2$ [10].

2. COULOMB: For a Coulomb potential $V(x) = -K/x$, Eq.(4) gives:

$$E_n = -K^2/[2(an)^2], \qquad (8)$$

which is *exactly* the usual functional dependence of the eigenvalues with the quantum number, reproducing all the known eigenvalues with $a = 1$ [10].

3. LINEAR: For a linear potential, $V(x) = Kx$, Eq.(4) gives:

$$E_n = \frac{3}{2}[K(an+b+c/n+\mathcal{O}(1/n^2))/2]^{\frac{2}{3}}, \qquad (9)$$

which is *exactly* the usual functional dependence of the eigenvalues with the quantum number, reproducing all the known eigenvalues up to four significant digits with $a =$

$1.81425$, $b = 0.45619$, and $c = 0.01803$ while stopping at $c$ [10], where the exact solutions are the zeroes of the Airy function [12].

4. RYDBERG: : For a Rydberg state, $n >> 1$ and $E_n = an$, and the potential for the motion in the $z = 0$ plane, in cylindrical coordinates [11], Eq.(4) gives:

$$V(\rho) = -\rho^{-1} + \frac{1}{2}T\rho^{-2} + \frac{1}{8}\omega^2\rho^2, \quad (10)$$

where $T = m^2 - 1/4$. Trivially,

$$E_n = \frac{1}{a^2n^2}[(\frac{1}{2} + \frac{T}{2n^2a^2})b^2 - b + \frac{a^6}{8b^2}\beta^2], \quad (11)$$

where $b = an\alpha$ is the positive root of

$$(1 + \frac{T}{a^2n^2})b^4 - b^3 - \frac{1}{4}a^6\beta^2 = 0, \quad (12)$$

with $\beta = n^3\omega$ [$n$ being the principle quantum number, $\omega = \frac{eB}{Mc}$ the cyclotron frequency, and

19

$B$ the magnetic field strength], which is *exactly* the usual functional dependence of the eigenvalues with the quantum number, reproducing all the known Rydberg state eigenvalues.

Based on these potentials and the spectra of all the other potentials tested [8, 10], we expect that *exactly* the usual functional dependence of the eigenvalues of the TISE in the discrete Eq.(3) is to be obtained with the quantum number for a generic potential $V(x)$, and we expect to produce all the eigenvalues.

We conjecture that Eq.(3) is an *exact* discrete representation of the TISE, without using *any* implicit or explicit continuity.

Besides eliminating the requirements of *continu-ity* by not using the second-derivative in a TISE solution of $f(x)$, the TISE becomes discrete, and we can introduce the series $f_n(x)$ as the desired linear combination of functions — as harmonic oscillator solutions (Fourier), Sturm-Liouville solu-tions, or Euler-Lagrange solutions, as desired, and the fields no longer have modes of arbitrarily short wavelength or time.

The solutions $f_k(x)$ are piece-wise valid in a varia-tional scale accounted for by the potential gradient itself, as in Eq.(3), and together with the other di-mensions, they span all four-dimensions providing a discrete topology for relativistic spacetime.

This new framework includes:

(1) QM, with a discrete topology for relativistic spacetime;

(2) energy, with a quantum spacing $1, 2, 3, ...N$ for the energy levels;

(3) geometry, with real numbers [6] in a physical dual space; and

(4) quantum entanglement, with a self-similarity (fractal dimension), linking energy and geometry.

---

[6] Calculated without limitation as finite rational numbers, as explained

In this framework, one has a many-body-relativistic-QM and entanglement is included specifically, without continuity.

One can write Eq.(3) as,

$$T_{n-1} = \Lambda(T_n)T_n, \qquad (13)$$

$T_0 = \Lambda(T_1)T_1$, $T_1 = \Lambda(T_2)T_2,...,$ where $n = 1, 2, 3...N$, the parameter $\Lambda$ is called the "quantum rate", and one progresses from the initial, known value at the origin $\psi(0) = 0$ onward, to a known value at the end $\psi(\infty) = 0$ (where $\infty$ is defined as above).

Eq.(8) can be seen with the *same* propagation onward as in a logistic map [4, 5], connecting both spaces, as with a mirror positioned between an object and the image.

The usual logistic map can be defined by the following equation:

$$x_{n+1} = \lambda_n x_n \qquad (14)$$

$x_2 = \lambda_1 x_1$, $x_3 = \lambda_2 x_2$,..., with $\lambda = [0...1]$ indexed to the non-negative integer numbers in $\lambda_n$, where 1 is the maximum rate, and is called the "growth rate", and one progresses from the initial, known value at the origin $x(0) = 0$ onward, to a value at the end $N$, taken as $x(N) = 0$.

It is known that this is a simple mathematical model with very complicated dynamics and chaos behavior [4].

The two maps in Eq.(8) and Eq.(9) are thus related by the same boundary conditions at the extremities, and steps onward, requiring only that $\Lambda_n$ is equivalent to $\lambda_n$.

Since both $\lambda_n$ and $\Lambda_n$ use boundary conditions at the start and at the end (both equal to zero), that must be satisfied, we expect that a quantum system ($\Lambda_n$) will be representable by one or more $\lambda_n$ systems, and vice-versa.

To confirm this hypothesis, we can, e.g., use the harmonic potential, previously calculated in Eq.(7). Conversely, we can synthesize a function in $\lambda_n$ that would obey an interesting behavior, such as bifurcation and chaos, and enquire about the quantum version in $\Lambda_n$.

This can be seen as the same propagation onward in a logistic map in chaos theory, connecting both spaces. This offers a quantum origin for the Feigenbaum constant.

The graph of a typical logistic map solution is shown in Fig.(1), which is akin to the quantum solution for the lowest eigenvalue in the time-independent Schrödinger equation.

One expects to be able to use the results with the logistic map to explain their origins in QM. For ex-

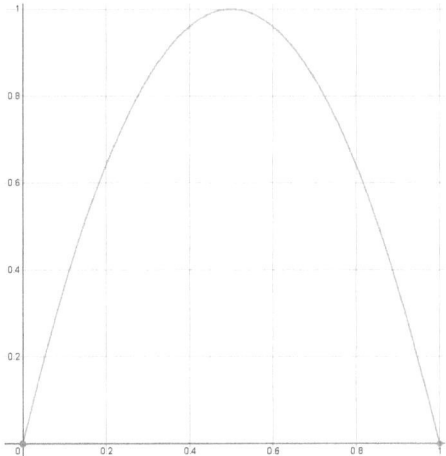

FIG. 1. Graph of a typical logistic map solution.

ample, the Feigenbaum constant, which seems to hold for a number of different logistic maps [5], for functions approaching chaos via period doubling.

## III. CONCLUSION

If the mathematical field is an abstract space in a choice of topology, we show here that the conventional choice of topology implies that the physical field of the standard QM is still utterly classical, and continuous, even if relativistic.

Quantum mechanics is well-known to be in a total conceptual rupture with classical mechanics, and with the concepts of continuity, sub-summed here as the many aspects and needs artificially imposed by *'relativistic field-continuity, uniformly varying variables (uniformity), local causality, infinity, $(\epsilon - \delta)$ Cauchy formalism, determinism, and renormalization'*, (here called figuratively the '7

Plagues' in physics), that we eschew and eliminate therefore the requirements that give origin to the '7 Plagues', so they cannot be pursued in a consistent theory.

We show how to bridge this limitation by using a first-order *difference* equation in discrete mathematics, that then leads to chaos theory and fractals [4, 5, 13–16], and help create the described framework, for a consistent, discrete quantum field theory (DQFT) without renormalization considerations.

Using discrete mathematics we achieve perfect accuracy and highest possible universality in a quantum universe with the DQFT, rather than postulating a continuity with so-called "real numbers", for example in a conventional QFT. All that we can measure are finite rationals, quantum, which are then finally mapped 1:1 to all that we can finitely count, quantum as well [17].

The DQFT can be reduced exactly to the time-independent Schrödinger equation for bound states in 3D, as shown, where all dimensions are treated on equal footing with a variational principle. With this equation, the energy field must have a discontinuous, discrete structure, without imposing continuity, in a natural way.

The DQFT is presented as the same propagation onward in logistic (iterated) maps in chaos theory, connecting both spaces, and offering a quantum origin for the Feigenbaum constant. Not using continuity preempts using renormalization and to explain universality in QM, and can be used to compute Feigenbaum's constants. This will be further extended elsewhere. Nature itself is discrete, as quantum, and finite. Well-suited for Galois fields as a topology of Nature's behavior in QM.

ACKNOWLEDGMENTS: The author is indebted to Edgardo V. Gerck and four anonymous reviewers. ResearchGate discussions were also used, for "live" feedback.

## Appendix A: The Quantum Mechanics Equations (QME)

Limitations in the accuracy of physics equations have everything to do with the mathematical assumption of continuity, which is false as we report, to solve them.

In standard QM, the Time-Independent Schrödinger Equation (TISE) [9, 11, 18, 19], albeit not considered in a Lorentz invariant form, is well-known. The TISE is one of only a few solvable models in QM, and shares many qualitative features with physically important models, e.g., quantum-well lasers [19] and Rydberg atoms [11]. Special relativity (SR) and Lorentz invariance were eventually added, as known, by the Dirac and Klein-Gordon equations [18].

But, the TISE, the Dirac, and the Klein-Gordon equations (here called QME, for quantum mechanics equations) are based on a troublesome, contradictory, underlying, field that is utterly continuous and uniform e.g., by requiring a second-derivative in a familiar expression of the TISE,

$$-\frac{\hbar^2}{2m}\frac{d^2\psi(x)}{dx^2} = [E - V(x)]\psi(x). \qquad (A1)$$

This work eliminates the continuity and other conditions, as unavoidably imposed in the continuous topology when considering the continuity, that are required for the second-derivative to exist [20] [7].

Without imposing thus an *ad hoc* continuity, this work will be able to better address the issues around precision, scalability, and computing, in quantum and computer code. Also, one does not need anymore to talk about imprecision when dealing with precision — instead, we talk about resolution (i.e., to separate or distinguish between closely adjacent objects).

The new formalism, discrete, and finite, is discussed here, along the context of the description of measurements and a now impossible decoher-

---

[7] A tangent plane remains continuous at the point, what we are pointing out is that *between* the points in a Galois field, the behavior does not have to be continuous.

ence in quantum theory. The avoidance of the Heisenberg uncertainty and observer effects was already presented in [21]. The special and general relativity harmonization with QM will also be further explored from [21].

To eliminate the requirement of continuity from the second-derivative in a QM setting for $f(x)$, so that the fields no longer have modes of arbitrarily short wavelength or time, we turn to a discrete representation – as the true representation of Nature, at the topological level.

Nature is then no longer seen to be continuous (as still used in [6]) and the digital is no longer an "imperfect" or "aproximate" measurement or representation of it, but the reverse. We posit that the Nature that we can measure (i.e., the Wirklichkeit) is indeed quantum, digital, and that "continuity"' is an illusion created by a universality run amok; whereas the usual assumption of "universality" [22] is considered [8].

The most general discrete representation is a (signed) finite rational, which is equivalent in a 1:1 map to a finite natural number [17]. This choice of topology is also abstract.

First, such field already complies with QM in a many-body framework.

---

[8] "Universality" says that many models with different properties at short distances can all give rise to the same predictions about the physics at long distances.

This motivates this work to use an experimentally-valid underlying discrete topology in SR at all scales, even abstract, that would represent a microscopic, discrete, quantum field in macroscopic interactions (even apparently violating "universality").

The choice of topology is given as any finite set of rational numbers in the set $\mathbb{Q}$. The space of $\mathbb{Q}$ is, in all cases of physical results, expressible as a real number $\mathbb{R}$ in a dual space at large distances (i.e., "universality"). The real numbers are thus not an expression of a reality that we can see and compute, which matches the assumption that nothing else can be measured in Nature [9]

In other words, the reality is given by $\mathbb{Q}$ exclusively, and $\mathbb{R}$ is at most a human representation of it. Immanuel Kant ascribes to Nature only a purely instrumental value. By contrast, we attribute to Nature a value that is independent from the judgment of a rational human.

Continuity is, therefore, not theoretically nor experimentally valid in this model. The use of derivatives, of any order, implies continuity.

This is experimentally flawed, as continuity does

---

[9] The mathematical Lebesgue measure does not apply to rational lengths in physics and is well-known not to be predictive of experiments.

not exist in the universe, and is mathematically deficient, as it does not solve the problem. Solutions are missing if derivatives are used, and pretend solutions are artificially introduced.

The idea emerges of a quantum/relativistic measure and yet, concurrently, a self-similar (of fractal dimension) measure, where fractal dimension is a generic term applicable to all the variants (cf. Mandelbrot) [14].

The quantum measure is, of course, identified with a finite set of natural numbers 1, 2, 3,... and the fractal dimension is identified with a finite set of rational numbers in $\mathbb{Q}$, through the dual space $X$ in the geometry.

However, the here-fictitious space of $\mathbb{R}$ is, in all cases of physical results, reducible to $\mathbb{Q}$ [21]. The importance of using $\mathbb{Q}$ instead of $\mathbb{R}$ is then not just as a "good approximation", or to be computationally faster, as noted in [21].

A set of finite natural numbers (that can be seen as a Galois field) are *equivalent* to a set of the finite rational numbers as a 1:1 map, and have the same cardinality, as defined by Cantor [17], which is used here to represent *quantum entanglement*.

As discussed in [21], this provides a connection between energy and geometry, through quantum entanglement, so that the flat/curved geometry of relativistic spacetime emerges naturally from the wave function.

### Appendix B: History

One hundred years ago, with Hubble, we already abandoned the idea that the universe would be infinite [10].

Earlier, starting with Poincaré [13], and later with the Heisenberg principle [23], observer and experiment seemed to not be resolved (as separable) in physics. There seemed also to be no objectivity in QM — as objectivity would be mind-independent, contradicting QM. But there is a coherent abstract

---

[10] What then led our language to accept for so long that infinity would exist? That things could, at the same time, be infinite and exist. One explanation is language itself. Much in the same way, water is tasteless to us because it is our major constituent.

view one can pursue in QM, which resurrects objectivity in an 'open reality' [24].

In 1956, Leon Brillouin [25] came to the conclusion
that the measurement of extremely small distances
is physically impossible, which preempts continuity.
We already presented in [21], in agreement with
Brillouin, showing new results including the scalability of quantum computing and the unification of
general relativity (GR) with QM.

Universality [22] and renormalization [26, 27] were
introduced, in the background scenario of derivatives and continuity, in quantum field theories
(QFTs) [28], in what we call "coping mechansms"
to deal with the non-sense produced by infinitesimals-

infinities, instead of eschewing them as clutches in the first place [25].

Infinity can be used as a symbol, meaning an unknown number in algebra, that is high enough as needed, which is not pre-determined, fixed (i.e., a number), but is a finite result that cannot be counted. In that way, infinity should be treated as a *finite irrational number*.

That said, we should replace all derivatives with corresponding difference equations, valid in terms of Galois fields. Cantor showed it first, albeit his result for infinity holds, when infinity is understood as above, a symbol for a high enough number, uncountable, and unknown, but definitely finite.

For example, there are many formulas for the first-order difference equation, in seeking a proper substitute for the first-derivative – but the forward difference is known to be unstable and should not be used [6].

By assuming a solution space, like the set U defined above in Sec.(2), that obeys the boundary conditions, even the second-order difference equation can be exactly expressed in that space by less points (e.g., 2) than expected (e.g., 3).

The use of a solution space that is Euclidean will generally require more points than the exemplified set U. It is very important if knowledge of the problem can be used to restrict the solution space before the problem is solved.

Therefore, all derivative formulas are transformable to difference equations, with proper care, and the solution space enters the picture, e.g., through the known boundary conditions (that do not have to be fixed, but are known).

In that, Brillouin already concluded that infinitesimals did not exist, and published, in 1956. It took about 30 years until we showed in [8–11] that, instead of calculating on an Euclidean tangential reference frame using second derivatives, one could *exactly* reduce the problem to a non-Euclidean frame that is piece-wise conforming with the quantum equation topology, and rewrite the second derivative *exactly* as a difference equation in just *two* points [8], then solve the equation and do better than conventional methods could, using the second derivative, and much quicker.

Then, it took another 30 years to today to conclude provably here that this is, in fact, a general method and that it works for *one* reason: *Nature itself is discrete, as quantum, and finite.*

Instead of artificially bring a problem to the Euclidean tangential plane and solve it with assuming infinitesimals-infinities and continuity, we showed that we could move to the problem's non-euclidean frame itself, and solve it naturally using its own frame, which Nature already set as a quantum topology, and solve it better and quicker. Fast Fourier Transforms (FFT) [6] did that earlier to Fourier transforms, as well-known, treating a signal as discrete, quantum, finite — not analogue — and the rest is history. Progress seems to be slow, as it took some 100 years to eschew infinity-continuity itself, but we do not use it anymore in physics, as Brian Greene explained, using renormalization in physics – which saved QED – and is much the same thing, in the 1950s.

Now, in 2020, putting it all together, we finally can confirm Brillouin in 1956, renormalization and Feynman in the 1950s, and throw away the clutches of infinitesimals, continuity, infinity as usually defined, and the 7 Plagues of physics and mathematics with it, and advance another notch.

The works cited treats it hopefully better and better in this grand picture. We unified chaos theory, fractals, folding in isomorphism, entanglement, special and general relativity, with quantum mechanics, fulfilling the visions of Poincaré and Brillouin from 100 years ago, and presenting new constructive bases for mathematics in analysis.

Progress seems slow and sometimes even painful, as constructive mathematics can tell, but pain is not suffering, and progress is worthy of even more pain if more would be needed.

## Appendix C: Galois Fields in Physics

We can now give the grand picture, as parts have been delivered. Thanks for your patience.

We can use all four operations $(+ - x \div)$ in the field of reals; one can also define derivatives and integrals by hypothesizing continuity (which brings along the infinitesimals and Cauchy epsilon-deltas).

This brings problems as Leon Brillouin noted in 1956, because nothing is an infinitesimal in Nature – thus, continuity is bogus as well as Cauchy epsilon-deltas, and renormalization in QFT, and whatever depends on them. Lots of work can now be recycled, both in mathematics and physics.

To solve this issue, one can turn to the idea of Galois fields in mathematics, to do the same four operations $(+ - x \div)$ in the *finite integers*, which is a field, and therefore supports them, but now precisely. There is no more 0.99... problem, or of $1/3 = 0.33....$ Goodbye continuity, in math and physics.

Every calculation is exact and faster when using Galois fields, as we do for even very large numbers in cryptography. For example, the encryption algorithm AES uses two bits, folded into 128 positions ($2^3$ elements). This has been used to propose exact and faster quantum mechanics in [21], where decoherence is resolved (separated), and entanglement uses fractals over finite rationals to link energy with geometry.

Still, derivatives and integrals (calculus) seem to be missing, and the derivations of characteristic polynomials in Galois fields do not seem to help. Instead, one can use the discrete math formulas in a desired space of functions, like used in here, for the second-derivative in the wave equation. Other n-order-derivatives can use the same methods to see them as differences, using the same space U of Sec.(2) or other, such as $E = 1, x, x^2, x^3$. Physics calculations can become exact and faster, and the speed of light in vacuum is already an integer, exact.

This is a move to code better and faster as well, as we can now do exact and faster calculations everywhere, using Galois fields, including calculus. Eventually, analysis will be studied better in Galois fields as well.

---

[1] E.P Hubble. A relation between distance and radial velocity among extra-galactic nebulae. Publ. Nat. Acad. Sci., 15, 168, 1929.

[2] Wendy L. Freedman, et. al. Final Results from the Hubble Space Telescope Key Project to Measure the Hubble Constant. The Astrophysical Journal, Volume 553, Issue 1, pp. 47-72., 2001.

[3] Stephen Hawking. A Brief History of Time. New York Bantam Books, 1998.

[4] R. May. Simple mathematical models with very complicated dynamics. Nature 261, 459–467., 1976.

[5] M. J. Feigenbaum. Universality in complex discrete dynamics. Los Alamos Theoretical Division Annual Report 1975-1976., 1976.

[6] Alan V. Oppenheim, Ronald W. Schafer. Digital Signal Processing. Pearson; 1 edition, 1975.

[7] James Plank, Kevin Greenan, Ethan L. Miller. Screaming Fast Galois Field Arithmetic Using Intel SIMD Extensions. Proceedings of the 11th Conference on File and Storage Systems (FAST 2013), 2013.

[8] E. Gerck. The Exponential Difference. Private communication, cited in report number: EAV-12/78, Laboratorio de Estudos Avancados, IAE, CTA, S. J. Campos, SP, Brazil. Copy online at `https://www.researchgate.net/publication/286625459_Matrix-Variational_Method_An_Efficient_Approach_to_Bound_State_Eigenproblems`, 1978.

[9] Augusto. B. d'Oliveira Ed Gerck, Jason A. C. Gallas. Solution of the Schrödinger equation for bound states in closed form. Physical Review A 26:1(1)., 1982.

[10] Jason A C Gallas Ed Gerck, A. B. d'Oliveira. New Approach to Calculate Bound State Eigenvalues. Revista Brasileira de Ensino de Física, 13(1):183-300, Jan 83., 1983.

[11] Robert F. O'Connell Jason A. C .Gallas, Ed Gerck. Scaling Laws for Rydberg Atoms in Magnetic Fields, 1983.

[12] C. Guigg and J. L. Rosner. Quantum mechanics with applications to Quarkonium. Phys. Reports, Vol 56, p. 169, Elsevier, 1979.

[13] Poincare, H. Sur Le Probleme Des Trois Corps Et Les Equations De La Dynamique. Acta Mathematica, Vol. 13, Online at http://www.mittag-leffler.se/sites/default/files/final_memoir.pdf, 1890.

[14] Gerald Edgar. Measure, Topology, and Fractal Geometry. Springer Verlag. p. 7. ISBN 978-0-387-74749-1, 2007.

[15] A. Libchaber, C. Laroche, and Stephan Fauve. Period doubling cascade in mercury, a quantitative measurement. Rev. Mod. Phys. 47 (4): 773., 1982.

[16] J. P. Gollub, H. L. Swinney. Onset of turbulence in a rotating fluid. Physical Review Letters. 35 (14): 927–930., 1975.

[17] Georg Cantor. Ueber eine elementare Frage der Mannig-faltigkeitslehre. Jahresbericht der Deutschen Mathematiker-Vereinigung., 1891.

[18] Eugen Merzbacher. Quantum Mechanics. John Wiley and Sons, Inc., 1970.

[19] Ed Gerck, Luiz Miranda. Quantum well lasers tunable by long wavelength radiation. Applied Physics Letters 44(9):837 - 839., 1984.

[20] Courant R., Hilbert, D. Methods of Mathematical Physics, Volume 1. Wiley, New York, 1989.

[21] Ed Gerck. Quantum Computer Scalability: Measurements and Decoherence in Quantum Theory. preprint, Research-Gate 340133193, 2020.

[22] John Preskill. Simulating quantum field theory with a quantum computer. PoS(LATTICE2018)024, 2018.

[23] W. Heisenberg. Über den anschaulichen Inhalt der quantentheoretischen Kinematik und Mechanik. Zeit. für Physik, 43 (3–4): 172–198., 1927.

[24] B. d'Espagnat. Quantum Physics and Reality. *Foundations of Physics*, 41:1703–1716, November 2011.

[25] Leon Brillouin. Science and Information Theory. Academic Press, N. Y., 1956.

[26] K. G. Wilson. The renormalization group: Critical phenomena and the Kondo problem. J. Physique Lett. 43, 211-216., 1975.

[27] Brian Greene. The Elegant Universe: Superstrings, Hidden Dimensions, and the Quest for the Ultimate Theory. *American Journal of Physics*, 68, 2000.

[28] A. S. Wightman R. F. Streater. PCT, spin and statistics, and all that. Princeton University Press, 1964.

# Quantum Computer Scalability: Measurements and Decoherence in Quantum Theory

Abstract

The conclusion that the measurement of extremely small distances is physically impossible,
which conflicts with continuity, was already considered by Leon Brillouin ca. 1956, and now is
revisited in this work. Consequently, report new
results including the scalability of quantum computing and the unification of general relativity with
quantum mechanics. In the context of the description of measurements and decoherence in quantum theory, this work presents the avoidance of
the Heisenberg uncertainty and observer effects,
providing improved quantum computer scalability.
It is shown here that one can observe a quantum
system without producing an uncontrollable disturbance in the system. The spectra is discrete, while

1

the conventional Heisenberg principle and observer effects are denied.

## I. INTRODUCTION

In standard QM, the Time-Independent Schrödinger Equation (TISE) [1–4], albeit not in a recognized Lorentz invariant form, is well-known. The TISE is considered one of only a few solvable models in QM, and shares many qualitative features with physically important models, e.g., quantum-well lasers [2] and Rydberg atoms [4]. Special relativity (SR) and Lorentz invariance were eventually added, as known, by the Dirac and Klein-Gordon equations [1].

By requiring a second-derivative as in the familiar expression by Schrödinger,

$$-\frac{\hbar^2}{2m}\frac{d^2\psi(x)}{dx^2} = [E - V(x)]\psi(x), \qquad (1)$$

3

the TISE, as well as the Dirac, the Klein-Gordon, the SR, and the general relativity (GR) equations, are all based on a troublesome, underlying, field that is utterly continuous.

The experimental impossibility of the measurement of infinitesimally small distances, the basis of the assumptions of continuity and uniformity in physics, was a problem already considered by Leon Brillouin [5] [1].

---

[1] Brillouin wrote, "An interesting outcome of this discussion in the conclusion that the measurement of extremely small distances is physically impossible. The mathematician defines the infinitely small, but the physicist is absolutely unable to measure it, and it represents a pure abstraction with no physical meaning. If we adopt the operational viewpoint, we should decide to eliminate the infinitely small from physical theories, but unfortunately, we have no idea how to achieve such a program."

*This work proposes how to solve the problem described by Brillouin, and achieves new results here reported, including the scalability of quantum computing and the unification of GR with QM.*

The TISE was fundamental also in the 1950s, mainly due to Wightman [6], has received more attention recently due to quantum computation needs [7, 8], and is useful when formulating a more modern relativistic quantum field theory (not only as a continuity-based QFT), which we will call a discrete quantum field theory (DQFT).

According to Ozhigov [8], the most known example on which the drawbacks of the TISE become evident is the quantum computer, because the traditional methods are not applicable to the investigation of its scalability. The problem of quantum computer scalability represents factually the old question of the description of the measurements and decoherence in quantum theory [2].

It is feared, e.g., that one cannot observe a quantum system without producing an uncontrollable disturbance in the system (e.g., observer effects and Heisenberg uncertainty as further studied in this work) [7, 8].

Also, in 'particle in a box' and tunnelling, attention is often restricted to the TISE [1] which, in

---

[2] If one wants to use a quantum system to store and reliably process information, then one currently needs to keep that system nearly perfectly isolated from the outside world [7, 8].

one dimension, is given by,

$$-\frac{\hbar^2}{2m}\frac{d^2\psi(x)}{dx^2} = [E - V(x)]\psi(x), \qquad (2)$$

subject to the boundary conditions $\psi(0) = \psi(\infty) = 0$.

The TISE is important even when discussing a Lorentz invariant time-dependent Schrödinger equation problem as in [2], where the electron motion is in practice dominated by the oscillation of the quantum-well in the laser field and, consequently, sees a laser-dressed potential described by a TISE.

We noted [9], however, that the TISE can be seen as a Sturm-Liouville (S-L) form [10],

$$y'' \sim Q(x)y, \qquad (3)$$

providing additional insights into the TISE from the work already done with the S-L form and the Euler-Lagrange (E-L) equation. This is pursued further here, first dealing with a new, coherent physics framework for the TISE, including QM and SR.

Second, this work eliminates without loss of *generality or accuracy* the continuity condition still existing from the mathematics, even in an otherwise physics quantum model — as unavoidably imposed in Eqs.(1-2) when considering the continuity that is required for the second-derivative to exist.

Without imposing *ad hoc* a continuity, this work

will be able to better address the issues around quantum computer scalability. We do not talk anymore about imprecision when dealing with precision — instead, we talk about resolution (i.e., to separate or distinguish between closely adjacent objects).

The new formalism, discrete and Lorentz invariant, is discussed here, along the context of the description of measurements and decoherence in quantum theory, with avoidance of the Heisenberg uncertainty and observer effects. The continuity condition is further discussed in the Appendices A and B.

## II.   THE QUANTUM THEORY OF THE TISE

The idea emerges of a quantum/relativistic measure and yet, concurrently, a self-similar (of fractal
dimension) measure, where fractal dimension is
a generic term applicable to all the variants (cf.
Mandelbrot) [11].

The quantum measure is, of course, identified with
the natural numbers 1, 2, 3,... and the fractal dimension is at first identified with the real numbers,
through the dual space $X$ of real numbers $\mathbb{R}$ in the
geometry.

However, the space of $\mathbb{R}$ is, in all cases of physical
results, reducible to $\mathbb{Q}$ for the following reasons:

- The use of any real number experimentally as a signed rational number $sp/q$ is well-known in mathematics, computer code, and physics, such as [3, 4, 9], as analyzed later.

- As used in [3, 4, 9], it does not matter if the spacing of the points physically assumed in the dual space $X$ is fixed, random, varying, a rational, an integer, a tensor, a potential, or a real number, which can *all* be fitted with a general fractal dimension that is discrete, as a signed rational number $sp/q$; and

- For any real number, all that one can physically measure, i.e., in a physically-validated way, are the signed finite rationals, $sp/q$; and

- Without any loss of generality, one can use $\mathbb{Q}$

only; and

- Using $\mathbb{R}$ has no benefit computationally, while it may cause a burden thereto.

The importance of using $\mathbb{Q}$ instead of $\mathbb{R}$ is then not just as a "good approximation", or to be computationally faster.

The natural numbers are *equivalent* to the (signed) rational numbers in Digital Constructivism (DC) [3] as a 1:1 map, and have the same cardinality, as defined by Cantor [17].

---

[3] Digital Constructivism (DC) [12] follows Brouwer's constructivism [13, 14], together with type theory [15], and the Curry-Howard correspondence [16]. Brouwer's constructive mathematics has the strict interpretation of the phrase "there exists" as "we can construct" —— which matches, from the start, with physics [8].

As we show later, this provides a connection between energy and geometry, through quantum entanglement, inducing a quantization of flat spacetime in SR, through what we call a *self-similar equivalence*.

The foregoing can also induce a quantization of spacetime *with* gravity (i.e., in GR), but distinct for each case as a fractal dimension.

The new framework for the TISE includes:

(1) QM, with a discrete topology for spacetime;

(2) energy, with a quantum spacing 1, 2, 3,...;

(3) geometry, with real numbers [17] in a physical dual space; and

(4) quantum entanglement, with a self-similarity (fractal dimension), linking energy and geometry.

We now bring together the points above, and see that the flat geometry of spacetime, SR, can emerge naturally from the wave function.

The properties of the 1:1 mapping of natural numbers to signed rational numbers, are derived as:

$$SNM \Leftrightarrow SN/M,\qquad(4)$$

where SNM is the natural number that results from the concatenation of S, N, and M, where all three are also natural numbers, as first described by Cantor [17], albeit without using concatenation as here, and in a more complex method. Signed expressions are included, where S stands for the encoding of the sign value.

We find then that the signed rational numbers SN/M, can map 1:1 to the natural numbers formed by the concatenation of S, N, and M (see Eq.(4)), describing the same quantum progression 1, 2, 3 ,..., and emerging as an expression for the *quantum entanglement*, linking energy and geometry.

Therefore, one can use the signed rational numbers as the dual space X, and:

a) endow X with the nice mathematical properties of a finite field. A finite field is a finite
   set which is a field; this means that multi-

plication, addition, subtraction and division (excluding division by zero) are defined and satisfy the rules of arithmetic known as the field axioms. Here, the finite fields are the set of integers mod $p$ when $p$ is a prime number, and the finite rational numbers together with addition and multiplication which contains the finite integers and is contained in any field containing the integers. Therefore, we consider the finite field of the (signed) rational numbers as X (e.g., 0 is included, negative numbers, closure, commutative, associative, distributive, inverse, and self-similarity), whereas the Lebesgue measure does not apply to rational lengths and is well-known not to be predictive of experiments; and

b) use X as an exact experimental local quan-
tity to apply in the physics [4], because that
is all we can measure: finite rational numbers. Where the rational numbers are a dense
subset of the real numbers because every real
number either is a rational number or has a
rational number arbitrarily close to it; and/or

c) choose subsets of X, dense enough in a piece-
wise global approximation [3, 4]; and/or

d) use it as a general method (the MVM) to solve
the S-L differential equation [9]; and/or

e) use it to model a stretched string, which can
then be understood to be equivalent in vibration to a discrete set of decoupled harmonic
oscillators (i.e., Fourier quantum frequencies),

reproducing in combination a range of arbitrary sounds [18]; and/or

f) other cases under *quantum entanglement.*

A DQFT that includes the TISE as a special limit case, is expected to be developed from the framework described by the four points (1-4) above.

## III.   AVOIDANCE OF THE HEISENBERG PRINCIPLE

Today, even though against what [1] proposes, the simultaneous appearance of classical wave and particle aspects no longer can be defended in QM.

That is due to the observation that the particle number need not be conserved. The relativistic dispersion relation in SR, that $E^2 = c^2p^2 + m^2c^4$, implies that energy can be converted into particles and vice versa.

The only thing that exists in Nature, as when following DQFT, is quantum waves, not particles and not waves as in Fourier analysis (classical).

Further, there is only one case (not two or even three cases) of photon interference in the two-slit experiment, and that is the case that is often neglected — the quantum wave case [4].

The Heisenberg principle (perhaps surprisingly to

---

[4] See any two-slit experiment at low-intensity, where just one photon is in the apparatus at any time. Watch online `https://www.youtube.com/watch?v=GzbKb59my3U`

some) depends thus on continuity, which is not theoretically or experimentally valid.

In other words, if Nature is already discrete (i.e., only the energy quantum levels as 1, 2, 3, ... exist in Nature) at its topological foundation, one should not try to "quantize anything" but work to find quantization in a geometry emerging out of that natural quantum framework, so that it is already provided with no field-continuity.

The Heisenberg uncertainty principle in QM seems, thus, to have to be modified — as it is in contradiction with itself, based on continuity by using the pure mathematics of the Fourier transform.

Further, to show that one can measure without disturbing, let us just look at the universe, estimated 13.8 billion years old. Since the universe is finite, the number of stars is finite, made by finite actions in a finite duration, and never halted in 13.8 billion years. All was measured, nothing went out of the laws that govern each phase, and did not go into uncontrollable confusion at each any state. Further, there is a difference between random and chaos. What looks random to one, might be chaos to another — utterly predictable.

There is no "decoherence" in quantum theory. We talk about resolution (i.e., to separate or distinguish between closely adjacent objects), instead of a pretend continuity.

## IV.   AVOIDANCE OF OBSERVER EFFECTS

"What is the frequency of a single photon?"

At first, the answer seems obvious by using TISE, as $E/h = f$, where $E$ is the energy of the photon, $h$ is the Planck constant, and $f$ is the frequency of the photon. But then one realizes that the photon would have to be observed in infinite duration in order to have a single frequency (e.g., Fourier transform relation, as observer effects in TISE).

The probability doctrine of QM, as given by TISE, asserts that the *indetermination*, of which we have just given an example, is a property inherent in Nature, and not merely a profession of our temporary ignorance from which we expect to be relieved by a future better and more complete theory [1]. Such more complete theory, however, appears to be quantum field theory (DQFT). The observer effects in QM may then have to be reexamined.

Obviously, then, $E/h = f$ given by TISE is not the correct answer. The *indetermination* was artificial, caused by the assumption of an infinite extension.

But nothing is infinite in Nature. We cannot wait forever to measure a photon, and nothing can. The Universe itself, as a finite manifestation, would not exist − one still would be waiting. The answer is to realize that something is wrong with the QM picture of a photon and TISE. The frequency of a photon is defined by its physical conditions in a DQFT, not by itself.

And it is not described by a Fourier transform either, which is a *mathematically continuous* procedure — with the hypothesis of infinitely close frequencies — and should never be used to represent a discrete phenomena, or artifacts of the interpolation will inevitably appear.

Why is DQFT better than QM? The answer below is relevant here, as QM is revealed to be subjective but DQFT is shown as intersubjective. Unlike pure mathematics, however, it is not enough in physics to be subjective — as Nature must be obeyed.

QM is based on deep untruths, as revealed by Nature, in addition to the rather formalist easy-to-solve fact that QM is not combining the principles of Lorentz invariance (SR-Minkowski-Einstein).

The untruths include: (1) One needs to abandon the single-particle approach of QM (subjectivity), as particle number need not be conserved, because of the relativistic dispersion relation in SR, mentioned in Sec.(5); (2) This requires a multi-particle framework (intersubjectivity), a many-body interaction with SR included and uses QM, it is a many-body-relativistic-QM, not just QM; and (3) Unitarity (basically, preserving the inner product) and causality cannot be combined in a single-particle approach, requiring intersubjectivity.

DQFT solves these problems by using a different approach: (I) The fundamental entities are not
the particles, but the field, an abstract object that
penetrates spacetime; (II) Particles appear as the
vibrations of the field.

The physical model of the photon, for example,
is given as a vibration of the EM field, and follows a DQFT. Then, in DQFT, the frequency of
the photon does NOT depend on the photon itself
and only (that would be subjective), but on its
physical conditions in a many-body-relativistic-QM
(intersubjective), such as the boundary conditions
at infinity. Then, that intersubjectivity can attain

objectivity — e.g., by being equivalent when seen by comoving [5] observers, in different experiments, at differing spacetimes.

Thus "relativistic" — which means the same laws of physics govern all observers who comove [26] relative to one another, so there is no preferred inertial frame of reference.

There is no "decoherence" in quantum theory. We talk about resolution (i.e., to separate or distinguish between closely adjacent objects), instead of a pretend indetermination.

---

[5] Same place *and* velocity. Comoving objects are not necessarily identical, but are considered equivalent when they see the same wave-front.

## V.   CONCLUSION

This work introduces a QM/SR framework supporting TISE, but not in continuous mathematics (e.g., with derivatives) but using the discrete mathematics of DC, as a topology.

Instead of talking about imprecision when dealing with precision, we talk in Secs.(5-6) about resolution (i.e., to separate or distinguish between closely adjacent objects).

Because the quantum levels are usually dense, they can be approximated "at a distance" by a resulting "continuity". However, there is no "decoherence" in quantum theory.

As a consequence, and important to quantum computing, it is shown here that one can observe a quantum system without producing an uncontrollable disturbance in the system.

The spectra is *always* discrete, coherent, quantum, and the conventional Heisenberg principle and the observer effects are denied in that resolution (i.e., to separate or distinguish between closely adjacent objects), albeit re-affirmed "at a distance".

This also lifts the usual assumption of "universality" in physics [16], which is to ignore "infinities" because it was assumed that there is limited sensitivity of long-distance physics ("the infrared") to the underlying physics at very short distances ("the ultraviolet").

We now can properly study macroscopic phenomena which are *sensitive* to the microscopic properties at the atomic distance scale, such as the laser, or even abstract as the empty space (vacuum states, with no physical particles) — in anti-universality [16].

The new framework includes: (1) QM, with a discrete topology for spacetime; (2) energy, with a quantum spacing 1, 2, 3,... as high as one may want; (3) geometry, with real numbers [17] in a physical dual space; (4) quantum entanglement, with a fractal dimension (*self-similar equivalence*), linking energy and geometry.

Here, we show that if one uses a natural connection (i.e., the *self-similar equivalence*) between energy and geometry, through quantum entanglement, this allows the macroscopic geometry of spacetime to emerge naturally, from the microscopic or even abstract, wave function — as flat (SR) or curved (GR).

The work also includes a quantization of spacetime with gravity, in GR. This approach can then introduce QM first into the topology, then into SR, and finally into GR [6].

---

[6] It is important to remark that there is no other quantum spectrum than the natural numbers, without loss of generality, even for real numbers [17] as a physical dual space for the geometry.

ACKNOWLEDGMENTS: The author is indebted to Edgardo V. Gerck and four anonymous reviewers. ResearchGate discussions were also used, for "live" feedback.

## Appendix A: SR And The Non-Existence Of Continuity

In calculating the TISE [3], one encounters an eigenvalue problem involving the operator in a S-L differential equation form [9], which frees one from the more restricted context with the TISE only.

---

The quantum distance could depend on the boundary conditions, energy, as well as mass (e.g., the gravitational binding energy, which reduces the mass).

One can *then* see the S-L differential equation in terms of the E-L equation, and the E-L equation specifies the variational solution to $\delta I = 0$ for the functional [10, 19]. Note that $I$ is also seen here as a kinetic plus a potential energy term, giving the Hamiltonian picture, which can lead back to the TISE.

As a well-known method of finding (e.g.) a function by means of its integral transform, the time-independent Schrödinger equation for bound-states, or TISE, can be recast as an isoperimetric variational problem using the S-L form (ie., the *transform space*), whereby the E-L equation is now equivalent to the *original problem* called TISE

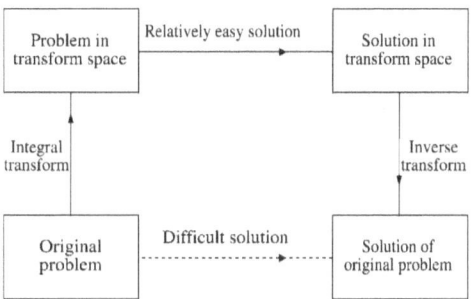

FIG. 1. Example of integral transform. From [20], page 934.

[10, 19], as shown in Fig.(1).

The idea in using here the S-L form is also to generalise the *academically continuous* (i.e., in pure mathematics) Fourier representation that is naturally associated to two related spaces, time and frequency, from *mathematical continuity* (Fourier representation) to a *discrete* sum of independent functions. E.g., as given in Appendix B at [18].

By requiring variations in all directions to equal zero, i.e., in $\delta I = 0$ above, the variational principle is equivalent to applying the E-L equation for each dimension in spacetime (preserving equal footing for space and time, as a Lorentz invariant solution). The Hamiltonian does not have to be introduced via the Lagrangian [21]. By using the E-L equation this way, however, one can derive the Klein-Gordon equation [1], achieving Lorentz invariance (SR). This is valid in SR as a flat spacetime, and in general relativity (GR) as a curved spacetime.

For example, this can be applied to expand a general function in terms of discrete sine functions, by making use of their completeness and their orthogonality properties, in order to connect the frequency and time domains.

It turns out that these properties are not unique to the trigonometrical functions expressed by discrete sine functions.

They can be regarded as following from solutions to a S-L problem, and any other S-L form equation will give rise to another set of functions with these completeness and orthogonality properties.

There is then, in general, for any S-L equation, a sequence of discrete eigenfunctions $y_n(x)$ that possess these completeness and orthogonality properties, such that a general function can then usefully be expanded as *discrete* $\sum_n c_n y_n$, in both spaces.

This happens not only as S-L solutions, but also as E-L solutions, and between spaces. Thus, it is possible to show [10, 19] that the discrete series

$$f(x) = \sum_i c_i u_i(x) \qquad (A1)$$

for a finite $n <= M$, where $M$ is as high a natural number as one wants, and with values taken from the E-L solutions, converges in the mean to $f(x)$, from a similar property shown for the Fourier transform, and S-L equation [10, 19].

An example of Eq.(3) in TISE is given in [3], where Eq.(1) of TISE is represented by the discrete expansion of $d^2\psi(x)/dx^2$ as $p\psi(x - q) + t\psi(x)$, in Eq.(2) of that reference, which is *exact* in the set $U$ of that paper, and allows many non-trivial linear combinations of functions in the set $U$, including piece-wise, that obey the boundary conditions.

This eliminates the requirement of *mathematical continuity* from the second-derivative in a TISE solution of $f(x)$, now discrete, and introduces its series as the desired linear combination of discrete functions $u_i(x)$ — as harmonic oscillator solutions (Fourier), S-L solutions, or E-L solutions, as desired, and the fields no longer have modes of arbitrarily short wavelength or time.

Using the now discrete TISE solution given by

Eq.(3), proposals using a discrete quantum field theory (DQFT), as a many-body-relativistic-QM, still pose some questions, reviewed next.

First, such unification between SR and QM in a many-body framework is considered necessary for the TISE, mainly because QM as given by the TISE does not follow SR, and is based on a troublesome, underlying, field that is utterly continuous (e.g., its use of the second derivative in Eqs.(1-2), which requires continuity), so the fields have modes of arbitrarily short wavelength or time.

The usual assumption [7, 22], for example in condensed matter physics and gravitation, is to to ignore such "infinities" because it is assumed that there is limited sensitivity of long-distance physics ("the infrared") to the underlying physics at very short distances ("the ultraviolet") as we typically study macroscopic phenomena which are insensitive to the microscopic properties at the atomic distance scale, or even abstract as the empty space (vacuum states, with no physical particles) — as the universality principle in physics [7].

This, however motivates this work to use an experimentally-valid underlying topology in SR at all scales, even abstract, that would represent an always discrete, quantum field everywhere, albeit being represented

---

[7] Universality says that many models with different properties at short distances can all give rise to the same predictions about the physics at long distances.

by a real number [8] in a dual space at large distances.

We then ask if the macroscopic geometry of spacetime (i.e., without mass, as in SR, or with mass, as in GR) can emerge naturally, microscopically or even abstractly, from the wave function.

### Appendix B: SR And QM

Continuity is, therefore, not theoretically-valid in QM — as derivatives require continuity. And continuity is known to be *incorrect* empirically in QM [9]. The lack of continuity could then be due either to: (1) our own (i.e., as an observer) inability to

---

[8] However, the space of $\mathbb{R}$ is, in all cases of physical results, reducible to $\mathbb{Q}$

[9] The quantum levels are not continuous; there is no 1/2 or fractional value, for example.   41

measure infinitely to zero; or (2) its non-existence.

To better distinguish the two issues, we verify that the assumption of continuity in SR comes from the very definition of arguments in the topology used, which is assumed to be continuous. But, the topology is always discrete in QM (one finds that no quantum process in Nature is continuous). It seems that one should use a discrete topology also in SR, in order to be compatible with QM. The conventional SR idea that a flat spacetime would be the manifestation of a spacetime that is *continuous* is certainly not experimentally-valid under QM.

We are also led by the positive fitting of other QM results:

(i) QED (quantum electrodynamics) as a successful QFT of photons,

(ii) QCD (quantum chromodynamics) as a successful QFT of gluons, and

(iii) SM (standard model) as a successful QFT where no particles or classical waves even exist, just quantum waves [23],

to conclude that, as the natural topology is experimentally discrete at many levels of description of Nature (i.e., QED, QCD, SM), and as "particles" are incompatible with quantum field theories [23], so Nature should also be discrete in the topology of SR.

The planets and stars thus, should not be made up

of hypothetical "particles", further adding to the physical reasons. Everything is QM in this picture, and a suitable *DQFT is expected to fit flat space-time, or SR.*

This shows that continuity should not exist in the TISE, including code [10]. A simple counter-experiment or counter-code can therefore prove that this is wrong, as a direct litmus test offered for its universal validity — just find (if one can) one counter-example with continuity in Nature including code.

This further says that continuity does not exist

---

[10] Such as in QM [8], biology, geology, computers, or AI, in past Nature (as we can see back for 13.8 billion years), and as in the Curry-Howard correspondence (code-as-proof).

in Nature [11].

---

[1] Eugen Merzbacher. Quantum Mechanics. John Wiley and Sons, Inc., 1970.

[2] Ed Gerck, Luiz Miranda. Quantum well lasers tunable by long wavelength radiation. Applied Physics Letters 44(9):837 - 839., 1984.

[3] Augusto. B. d'Oliveira Ed Gerck, Jason A. C. Gallas. Solution of the Schrödinger equation for bound states in closed form. Physical Review A 26:1(1)., 1982.

[4] Robert F. O'Connell Jason A. C .Gallas, Ed Gerck. Scaling Laws for Rydberg Atoms in Magnetic Fields, 1983.

[5] Leon Brillouin. Science and Information Theory. Academic Press, N. Y., 1956.

[6] A. S. Wightman R. F. Streater. PCT, spin and statistics, and all that. Princeton University Press, 1964.

[7] John Preskill. Simulating quantum field theory with a quan-

---

[11] Continuity also gives the wrong intuition.

tum computer. PoS(LATTICE2018)024, 2018.

[8] Y. I. Ozhigov. Constructive Physics (Physics Research and Technology). Nova Science Pub Inc; UK ed., ISBN 1612095534, 2011.

[9] Edgardo V. Gerck and Ed Gerck. On the Matrix-Variational Method (MVM) for Solving the Sturm-Liouville Differential Equation. Planalto Research, ISBN 1707639469, `https://www.amazon.com/dp/1707639469`, 2019.

[10] Courant R., Hilbert, D. Methods of Mathematical Physics, Volume 1. Wiley, New York, 1989.

[11] Gerald Edgar. Measure, Topology, and Fractal Geometry. Springer Verlag. p. 7. ISBN 978-0-387-74749-1, 2007.

[12] Ed Gerck. Mathematics Without Accidents, Part I. Planalto Research, ISBN 1700633929, `https://www.amazon.com/dp/1700633929`, 2019.

[13] L.E.J. Brouwer. Over de Grondslagen der Wiskunde. Doctoral Thesis, University of Amsterdam; reprinted with additional material, D. van Dalen (ed.), by Matematisch Centrum, Amsterdam, 1981.

[14] E.Bishop, E. and D. Bridges. Constructive Analysis. Grundlehren der mathematischen Wissenschaften, 279, Heidelberg: Springer Verlag, 1985.

[15] Institute for Advanced Study Univalent Foundations Program. Homotopy Type Theory: Univalent Foundations of Mathematics. Online at `https://homotopytypetheory. org/book/`, 2013.

[16] H. B. Curry. Functionality in Combinatory Logic. Proceedings of the National Academy of Sciences of the United States of America, Volume 20, Issue 11, pp.584-590, 1934.

[17] Georg Cantor. Ueber eine elementare Frage der Mannigfaltigkeitslehre. Jahresbericht der Deutschen Mathematiker-Vereinigung., 1891.

[18] Edgardo V. Gerck and Ed Gerck. New Physics With The Euler-Lagrange Equation: Going Beyond Newton: On-ramps to Quantum Mechanics, Special Relativity, and Noether Theorems. Planalto Research, ISBN 170438091X, `https://www.amazon.com/dp/170438091X`, 2019.

[19] Robert Weinstein. Calculus of Variations. Dover Publica-

tions, 1952.

[20] George B. Arfken. Mathematical Methods For Physicists. Elsevier Academic Press, 2005.

[21] Richard H. Price, Kip S. Thorne. Lagrangian vs Hamiltonian: The best approach to relativistic orbits. American Journal of Physics 86, 67, 2018.

[22] C. Misner, K. Thorne and J. Wheeler. Gravitation. W. H. Freeman and Company, ISBN 9780691177793, 1973.

[23] Art Hobson. There are no particles, there are only fields. Am. J. of Physics 81, 211, 2013.